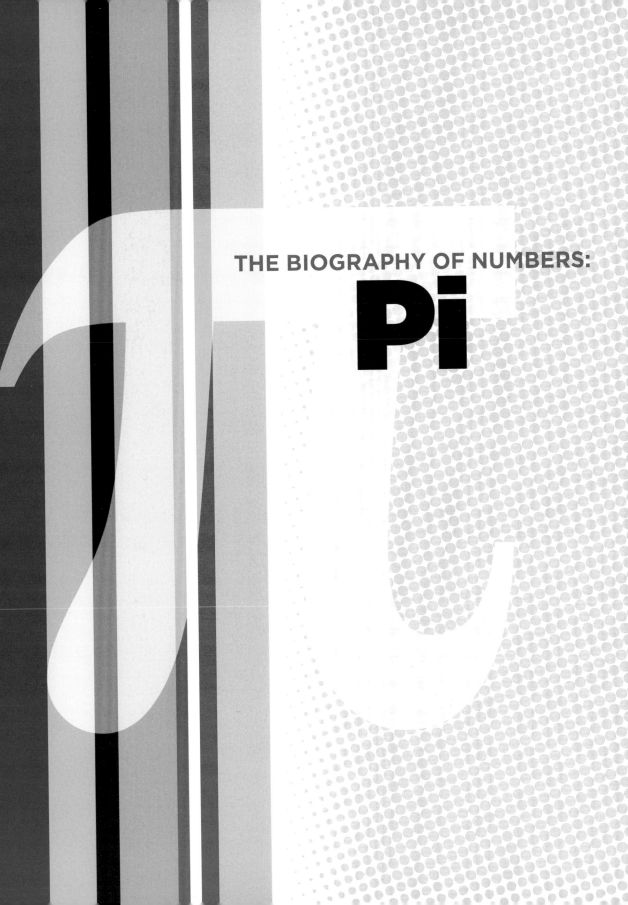

THE BIOGRAPHY OF NUMBERS:
Pi

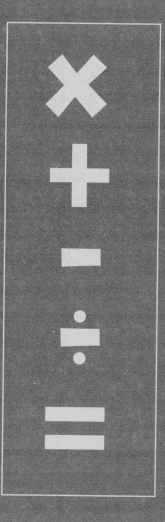

THE BIOGRAPHY OF NUMBERS:
Pi

Kevin Cunningham

MORGAN REYNOLDS
PUBLISHING

GREENSBORO, NORTH CAROLINA

To join the discussion about this title, please check out the Morgan Reynolds Readers Club on Facebook, or Like our company page to stay up to date on the latest Morgan Reynolds news!

THE BIOGRAPHY OF NUMBERS:
Pi

Library of Congress Cataloging-in-Publication Data

Cunningham, Kevin, 1966-
 Pi / by Kevin Cunningham. -- First edition.
    pages cm. -- (The biography of numbers)
 Includes bibliographical references and index.
 ISBN 978-1-59935-394-4 -- ISBN 978-1-59935-395-1
(e-book)
1. Pi. 2. Irrational numbers. I. Title.
 QA484.C86 2013
 516'.152--dc23
                                        2013011595

Printed in the United States of America
First Edition

Book cover and interior designed by:
Ed Morgan, navyblue design studio
Greensboro, NC

Contents

CHAPTER 1

The EARLY SEARCH for Pi

Mathematicians call pi a constant, a number with a fixed, never changing value. It represents what on the surface sounds unromantic, even dull. Pi is the **ratio** of the circumference of a circle (the distance around a circle's outside) to its diameter (the distance across from one side of a circle to the other). As pi never changes value, the ratio applies to every circle, of any size, everywhere.

The fascinating ratio

Just as pi never changes, it also never ends.

Most of us represent pi in its shorthand form of 3.14. But pi goes on for *trillions* of digits. Mathematicians have yet to find any pattern in the numbers. Not that they really need that many. Taking pi to just thirty-nine digits allows the calculation of a circle capable of holding the known universe.

Pi exists as a part of nature. Human beings did not invent pi. They discovered it, after a search that took thousands of years.

Though pi may sound of interest to only mathematicians, builders of circular rooms, and students in math class, it has fascinated human beings since ancient times.

The Reading Room of the British Library, now housed in the British Museum, is an example of a circular room.

Age old problem

An observer without much expertise in math could suspect pi's existence. Looking at a circle, perhaps measuring its diameter with a piece of string, might encourage a guess that the ratio was around three.

The earliest evidence of a civilization using a number near pi comes from a clay tablet dug up at Susa, a city in ancient Sumer. The tablet, dated from before 1680 BCE, suggests the Sumerians used a ratio of 3.125. Not exactly pi, but close enough for everyday use.

Another ancient document, the Rhind Papyrus, suggests that Egyptians around 1650 BCE used their own formula to arrive at 3.1605 as the ratio. Like their peers in Susa, the Egyptian mathematicians probably reached their conclusions by measuring rather than calculation.

The Rhind Papyrus

Pi

The Great Pyramid at Giza, constructed c. 2589–2566 BCE, is approximately equal to 2π or 6.2832. Based on this ratio, some Egyptologists concluded that the pyramid builders had knowledge of pi and deliberately designed the pyramid to incorporate the proportions of a circle. Others maintain that the suggested relationship to pi is merely a coincidence.

Squaring the circle

Ancient mathematicians in Egypt and elsewhere used geometry to investigate a pi-related problem known as **squaring the circle**. To square a circle, one had to draw a square using only a straightedge. Its area had to equal that of a circle drawn with a compass. Furthermore, the problem had to be solved in a **finite** number of steps.

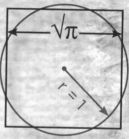

The Greeks picked up the challenge.

Anaxagoras

Born around 500 BCE, the philosopher-scientist Anaxagoras spent much of his career investigating natural phenomena like meteors and rainbows. Legend has it that local authorities imprisoned him for saying the sun was a gigantic hot stone rather than a god. While in jail, Anaxagoras supposedly used a straightedge and compass to tackle squaring the circle.

Anaxagoras failed. But other mathematicians kept trying.

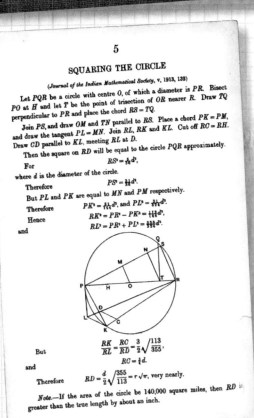

5

SQUARING THE CIRCLE

(Journal of the Indian Mathematical Society, v, 1913, 132)

Let PQR be a circle with centre O, of which a diameter is PR. Bisect PO at H and let T be the point of trisection of OR nearer R. Draw TQ perpendicular to PR and place the chord $RS = TQ$.

Join PS, and draw OM and TN parallel to RS. Place a chord $PK = PM$, and draw the tangent $PL = MN$. Join RL, RK and KL. Cut off $RC = RH$. Draw CD parallel to KL, meeting RL at D.

Then the square on RD will be equal to the circle PQR approximately.

For
$$RS^2 = \tfrac{5}{36}d^2,$$

where d is the diameter of the circle.

Therefore
$$PS^2 = \tfrac{31}{36}d^2.$$

But PL and PK are equal to MN and PM respectively.

Therefore
$$PK^2 = \tfrac{31}{144}d^2, \text{ and } PL^2 = \tfrac{31}{324}d^2.$$

Hence
$$RK^2 = PR^2 - PK^2 = \tfrac{113}{144}d^2,$$

and
$$RL^2 = PR^2 + PL^2 = \tfrac{355}{324}d^2.$$

But
$$\frac{RK}{RL} = \frac{RC}{RD} = \frac{3}{2}\sqrt{\frac{113}{355}},$$

and
$$RC = \tfrac{3}{4}d.$$

Therefore
$$RD = \frac{d}{2}\sqrt{\frac{355}{113}} = r\sqrt{\pi}, \text{ very nearly.}$$

Note.—If the area of the circle be 140,000 square miles, then RD is greater than the true length by about an inch.

Research article describing an approximate squaring of the circle using ruler and compass.

Antiphon of Athens proposed a method of finding the approximate area of shapes using **polygons**. Eudoxus refined Antiphon's work, later named the method of exhaustion, a technique used thereafter by Greek mathematicians to solve problems related to geometric figures, including circles.

The problem of squaring the circle, meanwhile, became so well known it entered popular culture. The popular playwright Aristophanes's work *The Birds*, written around 414 BCE, mentioned squaring the circle. It also gave the phrase its modern meaning of trying to accomplish a seemingly impossible task.

Aristophanes

Pi is referenced twice in the Bible. The First Book of Kings and Second Chronicles offer specifications for the Temple of Solomon, built by the great biblical king in about 950 BCE. *And he made a molten sea, ten cubits from the one brim to the other: it was round all about, and his height was five cubits: and a line of thirty cubits did compass it about.* The measurements show that Solomon's builders used the calculation of pi equals three.

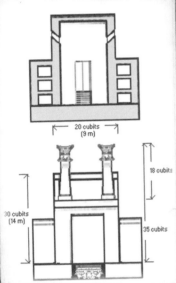

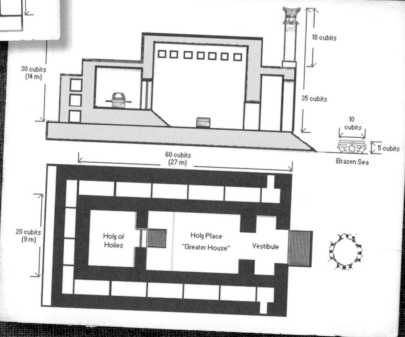

An artist's rendition of Solomon's Temple

CHAPTER 2

GENIUSES at WORK

Like other ancient peoples, the Greeks used three to approximate pi in nontechnical matters. But they recognized the need for precision in certain areas. One approximation placed pi at an inaccurate, but closer, 3.1622. One of the greatest scientists of the ancient world, however, would get closer still.

Archimedes

Archimedes was born in Syracuse, Sicily, then a Greek colony, about 287 BCE. Gifted in mathematics, astronomy, physics, and engineering, Archimedes's ideas and inventions aided Syracuse's prosperity. Archimedes had no fear of challenges. He once calculated how many grains of sand would fit inside the known universe.

3.1485

A sixteenth century statue by Simon-Louis Boquet depicting Archimedes

Polygons and pi

In his study of pi, Archimedes computed the area of two polygons with a formula known as the Pythagorean Theorem. The task represented an incredible feat of calculation. He started with simple six-sided hexagons and kept doubling the number of sides—recalculating each time—until his ninety-six-sided polygons *almost* equaled the area of the circle.

This new application of the method of exhaustion did not give him pi, and Archimedes knew it.

Instead, his work revealed that pi had to lay between the numbers $3\,^1/_7$ and $3\,^{10}/_{71}$. The numbers represented the upper and lower ranges represented by the polygons. The average between the numbers was 3.1485, the most accurate approximation of pi thus far.

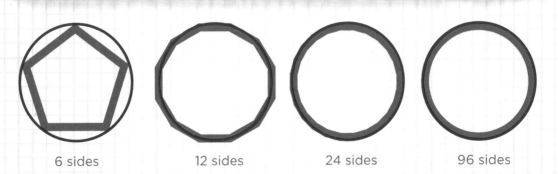

6 sides   12 sides   24 sides   96 sides

Archimedes calculated pi by doubling a hexagon until he
reached a ninety-six-sided polygon to find his estimate
which almost equaled a circle.

To four digits

The collapse of the Western Roman Empire starting in 476 CE began a period of intellectual stagnation in Europe. The small scientific advances that did take place did so within the monasteries of the Roman Catholic Church. Most of European society forgot the wealth of ancient knowledge.

Meanwhile, in Asia, the investigation into pi continued.

By 400 CE, India had entered a golden age of advancement and accomplishment. Mathematics, considered part of astronomy in India as it was elsewhere, played its part in the intellectual resurgence.

Aryabhata wrote his *Aryabhatiya*, the first of two major works, in 499 CE while still in his early twenties. Though the book largely concerned astronomy, Aryabhata also worked out pi to the correct 3.1416. (Note that in the sequence, the six is actually a five followed by a nine. Aryabhata rounded up.)

Historians do not know how Aryabhata calculated the figure. The *Aryabhatiya* did not show his work. But Aryabhata's discussion of the problem suggests he knew his figure was an approximation.

Aryabhata

Family business

Born to a family of astronomers around 430, Zu Chongzhi attracted attention for his intellect in young adulthood. As an adult, Zu independently came up with a method similar to Archimedes's. His polygons, however, had more than 12,000 sides. The huge number of calculations suggested by that figure took an immense amount of work. But it paid off. Zu stated pi fell between 3.1415926 and 3.1415927.

His fraction $^{355}/_{113}$ (called *Milu* or "detailed approximation") was accurate to six digits of pi. Mathematicians elsewhere would not calculate pi to six digits for a thousand years.

Zu Chongzhi

Zu achieved his impressive feat with a willingness to do the grueling and time-consuming calculations necessary to advance pi by even one digit. But his work did not involve advances in mathematical theories.

For the same thousand years, and beyond, calculating new digits of pi would demand stamina as well as intelligence.

3.1415927

Myth, Man and Mind

Archimedes, like other notable ancient figures, inspired many myths.

According to one story, he built a crane to operate a giant claw capable of seizing ships. Another held that Archimedes arrayed polished bronze or copper mirrors to focus sunlight on enemy ships and cause them to catch fire.

Archimedes's claw

Archimedes's real accomplishments were just as impressive. His pulley system allowed sailors to heft loads far beyond human strength. The Archimedes screw, assuming Archimedes really invented it, raised water from low lying areas into ditches that watered crops.

In 214 BCE, Rome attacked Syracuse. Archimedes's mechanical devices, including an advanced catapult and (perhaps) the giant claw, helped the city hold off the Romans until 212 BCE. Archimedes, then in his seventies, died during the Roman rampage through the city. A Roman soldier supposedly killed Archimedes when the great scientist refused to look up from doing a math problem.

CHAPTER 3

From
MORE
PLACES
to π

Starting in the eleventh century, increased contact with the Muslim world renewed interest in mathematics in Europe. But the next leap took place four hundred years later far to the east.

Beyond Zu

Born into poverty to Persian parents, Jamshid Masud al-Kashi reached adulthood as renewed interest in the sciences swept his homeland. Ulugh Beg, ruler of the local empire, founded a center for learning, called a madrasah, in Samarkand. He invited **scholars** from across the Muslim world to learn and teach in it.

Al-Kashi joined the migration of scholars to the madrasah. From the start he stood out as the pre-eminent practitioner of both astronomy and mathematics.

The Ulugh Beg Madrasah in modern times

Around 1424, he published his *Treatise on the Circumference*. In it, al-Kashi used polygons to take pi to sixteen decimal places. He also noted that a small error in computing pi becomes a larger error as the circles under investigation become larger.

Pi in Europe

Europeans rediscovered pi's mysteries, and much other knowledge, during the **Renaissance**. None knew of al-Kashi's work, however. It took years before a European equaled it.

In the meantime, Valentin Otto, an astronomer and thus a mathematician, independently discovered Zu's $^{355}/_{113}$ ratio in 1573. Frenchman Francois Viete practiced law and politics but also pursued mathematics and took pi to nine places using 393,216-sided polygons.

Ludolph van Ceulen combined even more exotic careers. In addition to teaching mathematics at the university in Leiden, Holland, he opened his own fencing school with himself as master. He managed these accomplishments despite lacking a college education. Nor did he speak Latin, the language of scholarship in his time.

Ludolph van Ceulen and his fencing school at the universitity in Leiden

Using Archimedes's methods, Van Ceulen computed pi to twenty decimal places. Work by him published after his death in 1610 carried it to thirty-five.

Europeans, especially in German-speaking lands, thereafter referred to pi as the "Ludolphine number" or "Ludolphine constant." Like Archimedes, Van Ceulen expressed his pi as a number between an upper and lower range of approximations. According to some sources, his family engraved the thirty-five digit Ludolphine number on his tombstone.

3.1415926
53589793
2384626
4338327
950289

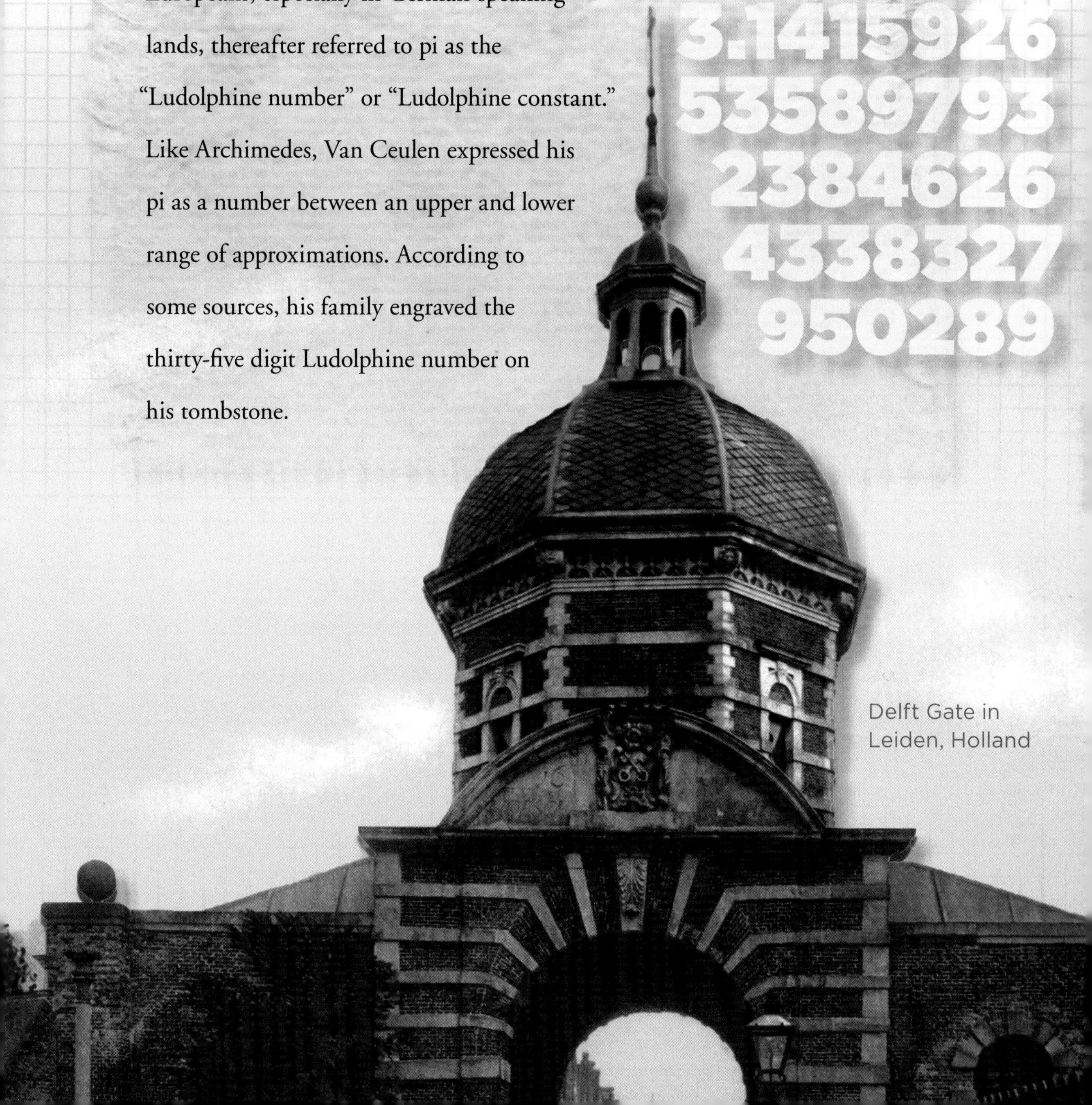

Delft Gate in
Leiden, Holland

Pi becomes π

English astronomer Abraham Sharp pushed pi to seventy-one digits in 1699. His countryman John Machin, using a new formula of his own creation, eclipsed Sharp's record by calculating pi to one hundred decimal places seven years later. The aptly named **Machin's Formula** became a favored tool for pi enthusiasts into the twentieth century.

$$\frac{\pi}{4} = 4\cot^{-1} 5 - \cot^{-1} 239$$

Abraham Sharp

The same year that Machin set a new record for calculating pi, a mathematician gave the ancient ratio a visual representation.

Before the 1700s, mathematicians did not use a symbol for pi. Instead, they referred to it in their writings with the unwieldy Latin phrase,

"[T]he quantity of which, when the diameter is multiplied by it, gives the circumference."

In equations, some used the letter p.

In 1706, William Jones, an English mathematician, adopted π (pronounced pi), the Greek letter for p, in one of his scientific papers. Though Jones became well known in mathematics, in 1706 he was just attracting attention. It took a bigger name to cement π as the symbol for pi.

The master

Capable of complex calculations in his head and blessed with a photographic memory, the Swiss mathematician-physicist Leonard Euler made important contributions in a wide number of fields. His system of mathematical notation remains in use today. Two important numbers, a series of equations, and formulas bear his name.

Leonard Euler

Euler used the shorthand "p" for pi early in his career. But in 1736 he switched to π in one of his books. What Euler did, others followed. His status as the greatest mathematician of his time brought π into the mainstream.

The π symbol gave mathematicians an easy bit of notation for working with the ratio. It also paid tribute to Archimedes and the other Greek thinkers who had launched the search for pi in Europe.

Van Ceulen's challenge

Ludolph van Ceulen's knowledge of pi brought him into conflict with a professor at his university in the 1590s. The professor had published a solution to pi that even Archimedes had known was wrong. Unable to criticize a powerful scholar, Van Ceulen quietly tried to convince the man's colleagues to convince him of his error.

He failed. The offended professor, unwilling to back off what he had written, challenged Van Ceulen to prove him wrong. Had the two men been evenly matched, it might have turned into one of the colorful feuds that crop up in scientific history. But as a junior scholar, Van Ceulen could not answer the challenge without endangering his career.

He triumphed nonetheless. His book *On the Circle*, published in 1596, made the professor's mistake clear and launched Van Ceulen into a distinguished career in mathematics.

This replica of the original tombstone of Van Ceulen at St. Peter's Church in Leiden, the Netherlands, is engraved with his thirty-five digit approximation to pi.

CHAPTER 4

The HUMAN ELEMENT

Van Ceulen had almost reached the thirty-nine decimal places of pi needed to calculate a universe-sized circle. Sharp and Machin went far beyond that mathematical necessity. But professional and amateur mathematicians continued to hunt for pi's next digits.

The geometry used for so long gave way to methods from trigonometry and the new discipline of **calculus** developed in the late 1600s. Both fields would provide formulas that opened up faster and more efficient ways of computing pi. At first, however, the digit count advanced slowly. In 1824 it had only reached 152, by British mathematician William Rutherford.

The pi hunters

The 1800s were the last century where only human minds, rather than humans teamed with computers, probed pi's mysteries. An unusual mix of personalities achieved some of the greatest successes.

Zacharias Dase was a **prodigy**. His talent: calculating numbers. Though gifted, Dase for many years had a hard time making a living. He had a modest education. Epilepsy, then untreatable, caused him ongoing health problems. He moved from job to job until in 1844 an Austrian professor

Zacharias Dase

Dase once calculated
79,532,853 x 93,758,479
in fifty-four seconds
(7,456,879,327,810,587)

suggested Dase use his talent for calculation on pi. Using a formula similar to Machin's, Dase took pi to two hundred decimal places in just over two months—an incredibly short amount of time.

Three years later, the Danish astronomer Thomas Clausen reached 248 digits.

William Shanks, an amateur enthusiast, far surpassed Clausen's achievement. In 1853 Shanks, working with Rutherford and employing Machin's formula, had published on pi. He revised his findings twice more, always correcting the errors he found. In 1873, he announced he had taken pi to 707 decimal places. Errors had crept into that tally as well, though it would be decades before anyone discovered them. But in the end Shanks was correct to 527 places.

The study of astronomy aided Clausen in his search for pi.

Unsquaring the circle

Immortality in mathematics was often related to brilliance. But it could be a product of timing or luck. Or, in the case of Charles Hermite, making one breakthrough and stopping short of another.

Hermite proved the existence of transcendental numbers, a concept important in algebra. A young German scholar named Ferdinand von Lindemann visited Hermite in Paris the year of Hermite's discovery. The two discussed Hermite's methods.

Charles Hermite

Paris in 1873

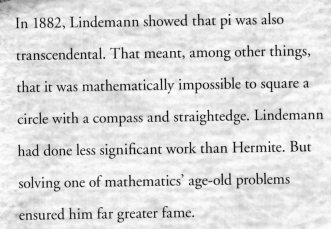

In 1882, Lindemann showed that pi was also transcendental. That meant, among other things, that it was mathematically impossible to square a circle with a compass and straightedge. Lindemann had done less significant work than Hermite. But solving one of mathematics' age-old problems ensured him far greater fame.

Ferdinand von Lindemann

Goodwin's pi

The 1800s was a time of incredible scientific advance. Pioneering research changed the understanding of biology, medicine, geology, genetics, chemistry, physics, and virtually every other field.

New ideas, however, bred resistance. Sometimes it came from established scientists who defended the traditional systems that had made their careers. In other cases religion played the major part. Others simply refused to accept change.

Hemming's Unicycle, advertised in 1869, was an example of the era's scientific advances. It had a wheel diameter of six to eight feet.

Indiana physician Edward J. Goodwin was an amateur mathematician who had long worked on squaring the circle. Refusing to believe Lindemann, Goodwin published an article in a mathematics journal claiming he had squared the circle, and that pi came out to 3.2.

Indiana law vs. scientific law

Goodwin copyrighted his finding in the belief that science and industry would have to pay to use his revolutionary work. But he did offer to let Indiana schools have the idea for free, with one catch. The government, he said, had to declare his "mathematical truth" of π=3.2. A legislator drew up a **bill** to make π=3.2 the law in the state of Indiana.

Goodwin's "truth" focused on squaring the circle. He appeared to have little interest in pi. The word *pi*, for instance, never appeared in the bill's language.

The legislation was heading toward a vote when Goodwin's maneuver became news. A professor at Purdue explained pi to confused legislators. The bill's chances soon faded. The Indiana Senate quietly, and indefinitely, put off the vote. Pi remained a matter of purely scientific law.

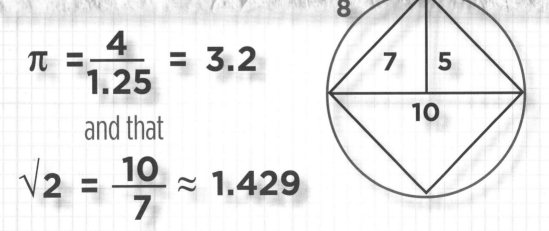

$$\pi = \frac{4}{1.25} = 3.2$$

and that

$$\sqrt{2} = \frac{10}{7} \approx 1.429$$

Goodwin's model circle as described in section 2 of the bill. It has a diameter of 10 and a stated circumference of "32" (not 31.4159~); the chord of 90° has length stated as "7" (not 7.0710~).

Buffon's needle problem

In 1777, the admired French naturalist Georges-Louis Leclerc, the Comte (Count) of Buffon, indulged his love of mathematics by proposing a problem. Say a person dropped a needle onto a floor (or table) made with strips of wood, with each strip the same width. What is the probability the needle will fall across a line between two of the strips?

Italian mathematician Mario Lazzarini answered the challenge in 1901. Lazzarini dropped a needle 3,408 times.

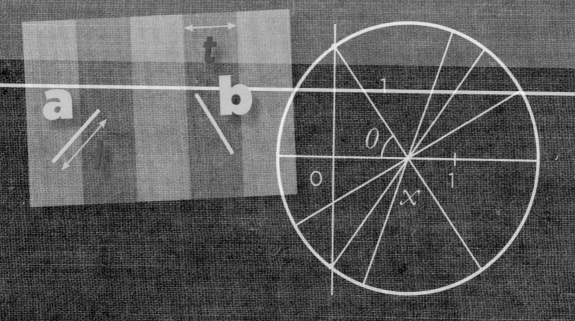

The result: a ratio of 3.1415929, or pi correct to six decimal places. Not all mathematicians accepted Lazzarini's findings in part because he used long needles with a greater chance of crossing a line.

The Comte of Buffon

Other experiments with a predetermined number of needles of a certain length have come up with a ratio of 3.116—not pi, but close to the rough estimates used in some ancient societies.

CHAPTER 5

CRUNCHING NUMBERS

Pi hunters had no noteworthy successes after Shanks claimed (wrongly, as it turned out) to have calculated pi to 707. His record stood well into the twentieth century. Shanks was not the last human pi champion. But his successors earned the distinction with the help of electronic calculators.

Shanks overthrown twice

D. F. Ferguson used a mechanical calculator to aid his pi hunting. In 1945, he discovered that Shanks had made a mistake at the 528th decimal place. The digit, and therefore all the digits that followed it, were wrong. Ferguson immediately surpassed Shanks's 527 correct decimal places by calculating pi to 710 places. Mathematicians John W. Wrench Jr. and Levi R. Smith confirmed, and then passed, Ferguson's numbers in 1947.

An early adding machine

ENIAC (Electronic Numerical Integrator And Computer) in Philadelphia, Pennsylvania

Then the computers took over.

Early computers were programmed with cards, filled whole rooms, and had virtually none of the capabilities we associate with present-day computers. But they calculated numbers at what must have seemed like a dizzying pace. ENIAC, the first digital computer, took pi to more than 2,000 digits in 1949. And did it in seventy hours.

Calculating pi became a test of a computer's power. In a short time, counting to hundreds of decimal places meant nothing. Not long after, counting to thousands or even millions, ironically, meant even less.

New era

Machine help offered no guarantee of success. A programming error, for example, foiled a 1957 attempt to reach 10,000 decimal points. The next year, an IBM computer managed the feat.

Wrench returned in 1961 with an IBM 7090 and collaborator Daniel Shanks (no relation to William). A noisy giant of a machine—more a roomtop than a desktop—the 7090 cost about $2.9 million and did work for the likes of NASA and the U.S. Air Force.

It also crunched numbers twice as fast as its predecessor. Armed with the state-of-the-art technology, Wrench and Shanks extended pi to 10,000 digits in under nine hours. In order to confirm the results, they performed two sets of calculations using different methods. Double checking became standard practice for computer-aided pi hunters.

IBM 7090 computers in a machine room at NASA during Project Mercury

Facing the limits

Computers provided invaluable aid. But the machines, like computers of any era, had limits on their calculating power.

In the 1960s, someone wanting to double the Wrench-Shanks record to 20,000 decimal places had to face that double the digits increased computing time by a factor of four, if not more. Even if a person thought he or she could program a computer to find, say, ten million places, it was not worth it to tie up a valuable machine—and computers were extremely rare resources—for the time it needed to do the calculations.

In 1973, for example, Jean Guilloud and Martin Bouyer used a fast computer of the time to calculate 1 million digits. Had they used their program to search out a billion digits, it would have taken the machine at least twenty-five years. And their machine could handle far more work than the 7090.

Taking pi further now depended on advances in computing power. More efficient programs took on a larger role, as well.

Jean Guilloud and Martin Bouyer used the CDC 7600 to calculate pi to 1 million digits.

Doubling down

Defined at a very simplified level, an **algorithm** is a procedure used to conduct a computer operation, like calculating pi. Algorithms that allowed for repeated (and repeated and repeated) efficient calculations took calculating pi to a new level.

Programmers and mathematicians at times drew on an earlier era for ideas. In the mid-1970s, Australian mathematician and computer scientist Richard P. Brent used work done in the early 1800s to create an algorithm able to find double the number of digits with each set of calculations. Eugene Salamin, another mathematician, found the same algorithm on his own.

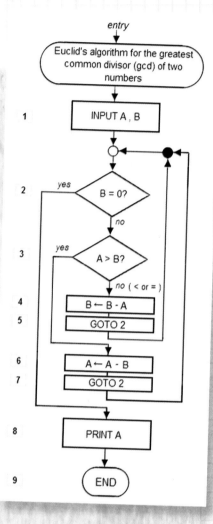

Euclid's algorithm shows a step by step calculation figuring the greatest common divisor of two numbers.

Because the Brent-Salamin algorithm uses huge amounts of computer memory, the Machin Formula remains the more common tool for pi hunters. But Brent and Salamin's work helped push our knowledge of pi into the billions of digits.

Racing for trillions

Mathematician Gregory Chudnovsky and his brother reached the 1 billion mark in 1989 using their own algorithms.

Over the next twenty-plus years, the Chudnovskys set several pi records. Often they did their work with supercomputers they built themselves in their apartment in New York City. The Chudnovskys stated their goal was not new records but to find out if the digits of pi followed a rule that distinguished pi from all other numbers.

Often the Chudnovskys traded the record with University of Tokyo professor Yasumasa Kanada. With colleague Yoshiaki Tamura, Kanada used a program that calculated 500,000 digits of pi every second to topple the Chudnovskys' record just three months after they set it.

In 1999, Kanada's team shattered its own record of more than 6 billion places by topping 51 billion digits of pi. Kanada led another team to more than 1 trillion digits, calculated in twenty-five days, in 2002. It took the team more than five years to write the computer program.

Kanada used a HITACHI SR8000/MPP supercomputer to compute pi to 1.2411 trillion digits.

Do it yourself

French computer programmer Fabrice Bellard set another pi record at the end of 2009. In 131 days, Bellard calculated pi to just short of 2.7 trillion digits. Bellard made his calculations not on a multimillion dollar supercomputer, but on a $3,000 home computer. Bellard, like the Chudnovskys, claimed he pursued pi for other goals—in his case, the development of algorithms with uses in other areas of computing.

Shiguro Kondo, a computer engineer for a Japanese food company, claimed the title for home-based number crunchers in 2010 (5 trillion places). Fourteen months later, he announced he had doubled his previous record.

Kondo worked on a self-built computer kept in his house. Since it ran nonstop, he paid whopping four hundred dollar per month electric bills. The machinery raised the room temperature to 104 degrees F., though his wife made it work by using the space to dry laundry. Kondo's effort survived the loss of a hard drive crash that wiped out two months' work. Later his daughter blew a fuse running her hair dryer. But a backup power source prevented a second disaster.

Persistent memory

Memorizing digits of pi became a contest after computers opened up the ratio to thousands of decimal places. Piphilology refers to the invention and use of memory techniques for remembering digits of pi.

Some practitioners write poems, nicknamed piems, that use the number of letters in a word to represent a digit of pi. As pi begins 3.14, the first word in a piem has three letters, the second one, and the third four. Ten-letter words represent zero.

3.14159265358979323846264338327950288419716939937510582097494459230781640628620899862803482534211706798214808651
3282306647093844609550582231725359408128481117450284102701938521105559644622948954930381964428810975665933446128
4756482337867831652712019091456485669234603486104543266482133936072602491412737245870066063155881748815209209628
2925409171536436789259036001133053054882046652138414695194151160943305727036575959195309218611738193261179310511
8548074462379962749567351885752724891227938183011949129833673362440656643086021394946395224737190702179860943702
7705392171762931767523846748184676694051320005681271452635608277857713427577896091736371787214684409012249534301
4654958537105079227968925892354201995611212902196086403441815981362977477130996051870721134999999837297804995105
9731732816096318595024459455346908302642522308253344685035261931188171010003137838752886587533208381420617177669
1473035982534904287554687311595628638823537875937519577818577805321712268066130019278766111959092164201989380952
5720106548586327886593615338182796823030195203530185296899577362259941389124972177528347913151557485724245415069
5950829533116861727855889075094838375415104944881526569604091627916710773847014940960165464023415204589353130237
0602491413427659052927802509464102922002...

(remainder of the decimal expansion continues)

Many champion piphilologists use an ancient memorization technique called the method of loci. With the method of loci, a person wanders a "memory palace" that associates an item (in this case a number) with specific details found in a familiar physical location like a house or a mall.

Akira Haraguchi, a sixty-year-old mental health counselor, recited 100,000 places of pi in 2006. Haraguchi memorized a series of self-penned stories and poems that help him remember the numbers in order. Haraguchi said his memory for all things improved after he started memorizing pi in 2001. Except, as he told a newspaper, he still forgets his wife's birthday.

Timeline

Biographical sketches

Archimedes (c. 285 BCE-c. 212 BCE)

Born in Syracuse on the island of Sicily, Archimedes became the greatest mathematician of ancient Europe, and also accomplished breakthroughs in engineering and astronomy. In addition to finding an approximation for pi, Archimedes learned to raise water from lowlands to irrigation ditches and designed weapons that defended Syracuse against Roman invasion.

Zu Chongzhi (429-500)

Zu Chongzhi learned astronomy and mathematics from family members in the field. His *Method of Interpolation*, written with his son, has been lost for centuries. But it is said the text included Zu's "detailed approximation" of $^{355}/_{113}$ for pi.

Aryabhata (476 CE-550)

Aryabhata worked in a number of areas of mathematics. In his text the *Aryabhatiya* (published c. 499), he calculated pi to four decimal places. Later in his career he concentrated on astronomy. In 1975, the government of India named its first artificial satellite Aryabhata.

Ludolph van Ceulen (1540-1610)

An expert fencer as well as a mathematics teacher, Van Ceulen spent much of his life investigating pi. After narrowly avoiding a pi-related feud with a superior at his university, Van Ceulen took the ratio to twenty decimal places in his 1596 book *On the Circle*. Work published after his death showed Van Ceulen later reached thirty-five places in his calculations.

John Machin (c. 1686-1751)

In 1706, the London astronomy professor John Machin won renown for calculating pi to one hundred places. But he became even better known for the formula he used, now called Machin's Formula, a method for calculating pi that remained a favorite of pi hunters well into the twentieth century.

William Shanks (1812-1882)

A boarding house owner and amateur mathematician, Shanks spent fifteen years calculating pi, and finally set a record by taking pi to 527 decimal places in 1873. (Shanks, unaware of an error in his calculations, claimed he had reached 707 places.)

Ferdinand von Lindemann (1852-1939)

A native of Hanover, now in Germany, Lindemann taught at the University of Freiberg. There, he proved that it was impossible to square a circle with a compass and straightedge, settling a problem that had vexed mathematicians for over 2,000 years.

Yasumasa Kanada (1948-)

A professor at the University of Tokyo, Kanada, along with his rivals the Chudnovsky Brothers, applied supercomputing technology to pi in the 1980s and beyond. A team led by Kanada broke the 1 trillion decimal place barrier in 2002, one of several records in his portfolio.

Glossary

π (pi)
The symbol π originally referred to the sixteenth letter of the Greek alphabet. Starting in 1736, π was generally accepted as the symbol used to represent the ratio pi.

algorithm
A procedure used to conduct a computer operation.

bill
In politics, a proposed piece of legislation.

calculus
The branch of mathematics that uses various functions to analyze change.

finite
Having a boundary or limit.

Machin's Formula
The mathematical formula used by John Machin in the early 1700s to efficiently calculate digits of pi, and used ever since for the same purpose.

Milu
The "detailed approximation" of pi by the Chinese mathematician Zu Chongzhi, expressed as $^{355}/_{113}$.

polygon
A closed two-dimensional shape made up of straight lines.

prodigy
A person with extraordinary talents or gifts.

ratio
The relationship between two numbers related in some way, as diameter and circumference in the ratio pi.

Renaissance (RE-ne-sahnts)
A period of European history beginning in the late 1300s and ending in the 1600s, marked by a revival of interest in ancient knowledge and great achievements in the sciences and the arts.

squaring the circle
An ancient mathematics problem of constructing a square with the same area as a circle in a finite number of steps with only a straightedge and compass.

Bibliography

Books

Beckmann, Petr. *A History of Pi*. New York: St. Martin's Press, 1976.

Berggren, J .L., Jonathan M. Borwein, and Peter Borwein. *Pi: A Source Book*. New York: Springer-Verlag, 2004.

Blatner, David. *The Joy of Pi*. New York: Walker & Company, 1999.

Gullberg, Jan. *Mathematics: From the Birth of Numbers*. New York: W. W. Norton, 1997.

Ifrah, George, and David Bello, trans. *The Universal History of Numbers*. New York: Wily, 2000.

Motz, Lloyd, and Jefferson Hane Weaver. *The Story of Mathematics*. New York: Basic Books, 1993.

Pickover, Clifford. *Archimedes to Hawking: Laws of Science and the Great Minds behind Them*. New York: Oxford University Press, 2008.

———. *The Math Book: From Pythagoras to the 57th Dimension*. New York: Sterling, 2012.

Posamentier, Alfred S., and Ingmar Lehmann. *Pi: A Biography of the World's Most Mysterious Number*. New York: Prometheus, 2004.

Smith, D. E. *History of Mathematics, Volume 1*. New York: Dover, 1958.

Periodicals and online

Aron, Jacob. "Epic pi quest sets 10 trillion digit record." October 19, 2011. *New Scientist*, October 19, 2011. http://www.newscientist.com/blogs/shortsharpscience/2011/10/pi-10-trillion.html.

Exploratorium. "A Brief History of Pi." http://www.exploratorium.edu/pi/history_of_pi/index.html.

McAvoy, Audrey. "Professor breaks own record—for the thrill of pi." Associated Press, December 6, 2002. http://www.seattlepi.com/national/article/Professor-breaks-own-record-for-thrill-of-pi-1102829.php.

National Public Radio. "Homemade computer sets records in the trillions." *Weekend Edition Sunday*, October 23, 2011. http://www.npr.org/2011/10/23/141629745/homemade-computer-sets-records-in-the-trillions.

Otake, Tomoko. "How can anyone remember 100,000 numbers?" *Japan Times*, December 17, 2006. http://www.japantimes.co.jp/life/2006/12/17/to-be-sorted/how-can-anyone-remember-100000-numbers/.

O'Connor, J. J., E. F. Robertson, and "Ludolph van Ceulen." The MacTutor History of Mathematics Archive, School of Mathematics and Statistics, University of St. Andrews. April, 2009. http://www-history.mcs.st-andrews.ac.uk/Biographies/Van_Ceulen.html.

Bibliography continued

————. "William Shanks." The MacTutor History of Mathematics Archive, School of Mathematics and Statistics, University of St. Andrews. July, 2007. http://www-history.mcs.st-and.ac.uk/Biographies/Shanks.html.

————. "Zacharias Dase." The MacTutor History of Mathematics Archive, School of Mathematics and Statistics, University of St. Andrews. April, 2009. http://www-history.mcs.st-and.ac.uk/Biographies/Dase.html.

Palmer, Jason. "Pi calculated to 'record number' of digits." British Broadcasting Corporation. January 6, 2010. http://news.bbc.co.uk/2/hi/technology/8442255.stm.

Preston, Richard. "The mountains of pi." *New Yorker*, March 2, 1992. http://www.newyorker.com/archive/1992/03/02/1992_03_02_036_TNY_CARDS_000362534.

Schudel, Matt. "Mathematician John W. Wrench Jr. dies at 97." *Washington Post*, March 25, 2009. http://www.washingtonpost.com/wp-dyn/content/article/2009/03/24/AR2009032403064.html.

Web sites

Buffon's Needle
http://www.mathsisfun.com/activity/buffons-needle.html
This math activity site shows how to conduct the Buffon's Needle experiment.

British Broadcasting Corporation
http://www.bbc.co.uk/programmes/p004y291
A group of professors discuss pi on a 2004 radio show broadcast in Great Britain. The site also offers shows on Archimedes, Indian mathematics, and other topics.

Exploratorium
http://www.exploratorium.edu/media/archive.php?project=80
The Exploratorium in San Francisco, California, provides pi-themed webcasts and videos, including a half hour introduction to pi and clips of Pi Day celebrations at the museum.

Public Broadcasting System
http://www.pbs.org/wgbh/nova/physics/approximating-pi.html
The science program *Nova* offers an interactive demonstration of how Archimedes conducted his experiment with polygons to determine pi.

"How to transform the number pi into a song."
http://www.npr.org/2011/03/14/134492882/how-to-transform-the-number-pi-into-a-song
An episode of a National Public Radio program features musicians who transform the digits of pi into music.

Index

Photo credits

All images used in this book that are not in the public domain are credited in the listing that follows: